I0489568

THE
CLIMATE CHANGE HOAX:

PATHWAY
TO
SOCIALISM

Lawrence Newman

The Climate Change Hoax: Pathway To Socialism

ISBN 978-1-7347100-5-2

Library of Congress Control Number: 2019905127

Second Edition

Publisher: Silver Millennium Publications, Inc.
          Gold Canyon, Arizona

Cover Photo: Solar eclipse, August 21,2017
             Goreville, Illinois

Other books by the author

Dedicated to all those who question the rush to judgment on man-made climate change

TO THE READER

The impetus for the writing of this book came to me while watching a total eclipse of the sun in Goreville, Illinois on August 21, 2017. As the disc of the moon moved over the sun's disc there was a noticeable cooling of the atmosphere on this hot day in southern Illinois. It struck me at that time, as I watched this magnificent celestial event, that global warming and cooling must be intimately involved with the fantastic power of the sun and not with some miniscule increase in the amount of carbon dioxide in the atmosphere that I had been reading about. I decided to do further research. I recommend that you do further reading on the subject as well. A list of recommended sources is shown at the end of this book.

I also urge anyone who has the chance to observe a total solar eclipse not to forego it. It's a visual event that will be burned in your memory forever. The next solar eclipse in the United States will occur on April 8, 2024.

"Eclipses are different (than other solar events). They truly fit the "amazing coincidence or divine plan" category. How else to explain that the moon is four hundred times smaller than the Sun but also four hundred times nearer to us? This makes the only two disks in our sky appear the same size. That would not be the case if either were larger, smaller, nearer, or farther away. Even more in line with "a divine plan" is the fact that since its creation four billion years ago from an interplanetary collision the moon has been slowly receding from

the Earth at the rate of 1- ½ inches a year. During the time of man's evolution on Earth the moon's face is currently in an optimum position that matches the face of the Sun."—Bob Berman from his book "The Sun's Heartbeat"

TABLE OF CONTENTS

CHAPTER 1

GENERAL COMMENTS—A QUICK OVERVIEW

Over the past three decades the public has been bombarded with the news that scientists have determined that humans are responsible for increased levels of carbon dioxide in the Earth's atmosphere. This increase, they claim, is leading to global warming, and that we must take drastic steps that will upend our lives in order to insure our survival. Furthermore, if we do not take the drastic steps many of these scaremongers predict, life on earth, as we know it, will be unalterably changed for the worse.

Many of us are left scratching our heads as the prophetic doomsday scenarios occurring in "just ten to twelve more years", which date back decades, keep failing to come to pass. However, these doomsayers keep pointing to hurricanes and other weather phenomena as proof of the coming apocalypse.

The term "scientific consensus" is bandied about with the implication that any doubters are knuckle-dragging Neanderthals. As long as these doomsayers forecasting the falling skies were scaring only each other and leaving the populace alone to conduct their lives they posed little danger. However, this has changed.

We are now faced with many major political figures endorsing the implementation of onerous regulations on our lives. Americans must be made aware how drastically their lives will change if these zealots of man-made climate change get their hands on political power. Hopefully, we will defeat these zealots at the ballot box. The people of France, "the yellow jackets", took to the streets against the elites when massive carbon taxes were imposed on the "little people", in the name of global warming. The elites, of course, would face little discomfort from these taxes as these flew around in their private jets and sped through the streets in their limousines.

The subject of global warming is a scientifically complex, highly politicized issue that requires intensive study by scientists and a reasoned approach as to the actions we can, or should, take. It should not be handled with scare tactics. The steps that we may take will have a very real effect on our energy consumption and how we and our children will live in the foreseeable future

After intensive reading about the subject of man-made global warming I came to the following conclusions:

- There may or may not be a global warming trend since the recent scientific studies on this subject by climatologists are not definitive. There's an old adage that says any result can be arrived at by the cherry-picking of facts. The climate prediction models developed by the climatologists continue to be proven wrong when compared to actual results. One of the insoluble problems with the development of climate change models is that cloud cover has a significant impact on ground temperatures and no climate change model has yet been developed that can correctly forecast ever-changing patterns of cloud cover.

- The impact of small increases in the amount of carbon dioxide in the Earth's atmosphere caused by human activity is minimal—if any. The probable primary source of any global warming is—obvious to most ordinary people—the Sun!

- The projected cost to reverse any global warming which may be occurring is prohibitive and would greatly reduce America's, and the world's, current standard of living. And most important, these huge expenditures will not prevent an unstoppable force.

- Our response to any global warming which may be occurring should be to adapt to it—not fight it.

- The historical record indicates that more people die from cold than from heat.

- The climatologists supporting the man-made theory of global warming are drinking from the trough of massive funding from many sources and have little reason to back off their claims. There are no government, business, university or private sources of funding for climatologists claiming that man is _not_ responsible for global warming.

- It is apparent that the man-made climate change hoax is based on a political, not a scientific foundation.

The chapters that follow will support the statements listed above and point out that the solutions to the man-made climate hoax are designed to force a socialistic government on the American people. What we are facing is simply a grab for political power by the elites using the global warming hoax as the basis for their proposed actions.

CHAPTER 2

THERE'LL BE A CHANGE IN THE WEATHER

Just as the opening line from a song of long ago forecasted a change in the weather that sentiment is just as true today.

The weather and the climate have been in a constant state of change for as long as humans have populated the Earth—and long before. The central question before us is whether the level of man's activities today are contributing in a major way to global warming. Climate scientists have tried to link the industrial activity, which began in earnest late in the nineteenth century, to an increase in the atmospheric carbon dioxide percentage. Many statistical climate models have been developed by climatologists attempting to link the increase in carbon dioxide levels to the warming of the planet. None of these climate prediction models has proven accurate. None! The complexity of the interaction between the forces influencing the weather, and the climate, are simply beyond the ability of today's climatologists to develop accurate predictive models.

A brief summary of facts regarding the forces that determine the temperatures and the climate on Earth would be useful.

Consider the following facts about the Earth we inhabit:

- The size of the Earth is ideal—if it were larger gravity would be too strong to allow life to exist as we know it; if it were smaller, the gravity would be too weak and the gases of the atmosphere would dissipate into space, producing an airless, desolate planet.

- The Earth is located in the "Goldilocks" zone of our solar system—a location outside this zone would have resulted in either a hot, arid planet or an icy ball.

- The rotation of the Earth is set to allow for the existence of a livable planet. If slower, it would have extreme temperatures; if faster, the atmosphere would produce violent winds, and extreme weather events. How the Earth attained this perfect time of rotation is an unanswered question better left to theologians.

- The combination of the Earth's rotation and the inner molten iron core produces an electromagnetic field, which protects the Earth from the solar wind and cosmic rays that would strip away the Earth's ozone layer and expose the Earth to harmful ultraviolet radiation from the Sun.

- The axis of the Earth allows the progression of seasons, and mitigates any extremes of temperature over much of the planet. Without the axis, which appears to be set at an ideal amount, the area close to the equator would be almost unlivable due to the heat and the southern and northern parts above the equator would have a very small habitable zone, with extremely low temperatures at the pole regions—far lower than today. Historical investigation by astronomers has come to the conclusion that the Earth was knocked onto its present axis by a collision with a Mars- size planet, early in the solar system's formation—the same collision that produced the moon. Why the Earth is at this exact ideal axis is also an unanswered question. Perhaps another subject best left to the theologians.

- The path of the Earth around the Sun is not perfectly circular. Sometimes it is closer than other times. This variation in distance from the Sun is caused by the gravitational pull of the other planets, primarily by the largest planet-- Jupiter. The distance from the Earth to the Sun ranges from approximately 94,450,000 miles to 91.340,000 miles and varies over a period of about 100,000 years, the approximate time period between Earth's warming and cooling cycles.

- The atmosphere of the Earth, excluding water vapor, is composed of 78% nitrogen, 21% oxygen, .9% argon and .1% other gases including carbon dioxide (.04%).

Consider the following facts about the Sun:

- The Sun at the center of our solar system is composed principally of the elements hydrogen and helium with small amounts of the heavier elements, oxygen, carbon, iron, neon and a few others.

- The source of the Sun's radiant energy is fusion, a complicated thermonuclear process occurring under tremendous pressure in the sun's core that converts two atoms of hydrogen into one atom of helium, releasing enormous amounts of energy as a result.

- The radiance (and the heat) of the Sun changes over time, for unexplained reasons, but appears to coincide with the emergence of sun spots.

"The bomb over Japan destroyed an entire city, simply from sucking several ounces of uranium out of existence, and transforming it into glowing energy. The reason the sun is so much more powerful is that it pumps 4 million tons of hydrogen into pure energy every second."—David Bodanis, from his book "E=mc²".

Lest the reader feel that the Sun will shortly burn itself up, since its hydrogen fuel is disappearing at such a rate, it is estimated that at the present rate of converting hydrogen to helium the Sun will maintain its current steady state of energy production for another 5 billion years.

The radiant power of the Sun travels to Earth in approximately 8 ¼ minutes, where it provides its life giving energy to warm the planet, create clouds and participate in the amazing process of photosynthesis which produces food for all living things on Earth.

Ninety-seven percent of the Earth's water is in the oceans and unusable by man. However, there is a miraculous system of recycling and purification in place. Evaporation takes the ocean waters into the atmosphere in fresh water form, leaving the salt behind. Earth's meteorological system then forms clouds, which moved by winds, caused by the Earth's rotation, distribute the water throughout the world for the use by plants, animals and people.

The major argument supporting the prediction of climate change is that the Earth is becoming a "hot house" due to an increase in greenhouse gases in its atmosphere. An increase in these gases are forecasted to prevent the heat of the Earth's surface from escaping into space. The greenhouse gas that the climatologists have labeled as the major culprit is carbon dioxide. A discussion of the importance of carbon dioxide to our existence will be taken up later. Suffice it to say at this point that carbon dioxide is a very minor factor in the total amount of greenhouse gas—which is 95% water vapor!

A study of changes in Earth's climate, done by the examination of long living tree rings, ice cores from the Arctic and Antarctic regions and man's historic records indicates major climate changes have occurred throughout history. Greenland, the largest island in the world, got its name during the time identified as "The Medieval Warming Period", hundreds of years ago, when the climate of Greenland supported areas of luxurious plant growth. Today, the island is mostly ice and snow covered. This period was long before the industrialization period that began late in the nineteenth century. This industrialization is blamed by the climatologists for the increase in carbon dioxide in the atmosphere.

One of the most interesting facts about the basis for the warning concerning global warming is the total disregard of the impact of the Sun. Most reasonable people would assume that the Sun is the major factor in any investigation of Earth's global warming. Paradoxically, the major organization pushing the climate change agenda has the following to say on this key point:

"The sun is an all-but-irrelevant factor in climate change."—conclusion of the report of the United Nation's Intergovernmental Panel on Climate Change.

The UN Panel commented further on this subject when it responded to an article in Discover magazine (July 2007) by Danish scientist Henrik Svenson., calling his findings "extremely naïve and irresponsible". Professor Svenson's scientific observations of the Sun showed that changes in the Sun's activity could explain most or all of the recent rise in the Earth's temperature.

This response by the UN Panel was expected since anyone not parroting the "scientific consensus" belief that climate change is caused by man's activities is considered heretical. If the UN Panel ruled during the days of the Spanish Inquisition Professor Svenson would have been burned at the stake.

Another scientist, questioning the UN Panel's conclusion, had this to say:

"Mars has global warming but without a greenhouse and without the participation of Martians"—Dr. Habibullo Abdusamatov, head of the Space Research Laboratory at St. Petersburg's Pulkovo Astronomical Observatory, is a strong believer that it is the Sun's cyclical changes in its irradiance which control the Earth's temperature, pointing out that the polar ice caps on Mars have been shrinking, in parallel with global warming on Earth.

The incomprehensible stand of the UN Climate Panel that ignores the power of the Sun on climate change would be laughable on its face, if it did not call for grossly exorbitant and unnecessary expenditures by developed countries throughout the world to halt an unstoppable force—monies that could be put to better use.

In an attempt to convince the public about the truth of man-made global warming scientists produced the "hockey stick" graph of average global warming over the past 1000 years. This graph shows a straight line for most of the period and then a significant jump beginning at the end of the 19th century, which coincides with the beginning of industrialization that produced significant amounts of carbon dioxide. This "hockey stick" graph was a prominent part of Al Gore's widespread movie presentation, "An Inconvenient Truth". Unfortunately for Mr. Gore and the scientists parroting his global warming scare tactics, the "hockey stick" and other so-called facts in his movie have been proven false.

However, in accordance with the old adage that a lie can travel around the world before truth gets its boots on, many people, especially the impressionable youth of developed countries, have been inculcated with the lies about man-made global warming. They have reacted in frightening ways. In addition to the introduction of the "Green New Deal" by youthful Congressional Representatives here in the United States there have been demonstrations in Europe.

In Britain, an organization naming itself the "Extinction Rebellion", has recently poured blood in front of 10 Downing Street, the British Prime Minister's official residence in London, camped out on Waterloo Bridge, stopping traffic, and chained themselves together in front of the Labor Party leader's home. The "Extinction Rebellion" has three demands:

- A declaration by the British government of a climate change emergency

- A reduction in greenhouse emissions to zero by 2025

- A formation of a Citizens' Assembly

All that seems to be missing with regard to the last demand, which has echoes of the Citizens' Assembly of the 18th century French Revolution, is the naming of a "Robespierre" to head the Assembly and the construction of a guillotine for climate change deniers.

One of the climate change zealots, David Roberts, a *Grist* magazine staff writer, has gone so far as to recommend Nuremburg-type trials for "these bastards" i. e. climate change deniers.

One other lie often quoted by the proponents of man-made climate change is that a "consensus of scientists" have subscribed to its belief. The number normally given is 97%. However, as is the usual case regarding outrageous claims, there is an interesting story behind this figure. The "97%" number was generated by surveying a select group of scientists who seemed inclined to support the theory of man-made climate change. In actuality there are literally hundreds of scientists who have gone on record denouncing this hoax. However their voices continue to be drowned out by the forces behind the climate change agenda.

In conclusion, there is no consensus of climatologists regarding the impact of man-made global warming. Instead there have been several studies which link the Sun's activity to changes in the Earth's climate. These studies have been totally disregarded by the UN Panel on Climate Change.   Instead the Panel continues to cling to its man-made climate change hoax and calls for actions that will cause major disruptions in the lives of people throughout the world.

CHAPTER 3

THE BUBBLES IN YOUR BEER

The attempt to label those carbon dioxide bubbles that rise up in your beer as a major greenhouse gas and, therefore the culprit changing the Earth's climate is a fool's errand. There are three very important facts regarding carbon dioxide that we all need to understand.

First of all, it is a very minor gas, representing 410 parts per million parts of the atmosphere, of which 5%. or only 20 parts per million is man-made. Consider a giant glass container of 1,000,000 ping pong balls. Further consider that 20 of these ping pong balls are black and the rest white. That's approximately one black ball for every 50,000 white balls. What do you think the chances are that you would see one of the black balls mixed into the glass container? The increase of carbon dioxide between 2005-2011 in the atmosphere measured by scientists was 2.21 parts per million, akin to a drop in the ocean. If we painted two of the 20 black ping pong balls red you would have some idea of the ludicrous nature of their potential impact on the climate. As stated earlier, water vapor is the major greenhouse gas—95%! Hopefully, there won't be any calls to eliminate water by the doom forecasting climate scientists.

Secondly, carbon dioxide is vital to life here on earth. Without it plant life would die. In fact, scientific studies have shown that the higher the carbon dioxide percentage in the atmosphere the better that plants grow, especially wheat and rice, two of the major grains feeding the world's population. Past periods in the Earth's evolution have shown much higher concentrations of carbon dioxide in the atmosphere—and luxurious plant growth as a result.

"Carbon dioxide is not a pollutant. Life on earth flourished for hundreds of millions of years at much higher carbon dioxide levels than we see today . . . cultivated plants grow better and are more resistant to drought at higher levels."—William Happer, professor of physics at Princeton University

The miracle of photosynthesis occurring in the green leaves of a plant, using sunlight and carbon dioxide, produces food for the plant and oxygen for the atmosphere—a gas vital for human existence. Photosynthesis is a very complicated process—one that is spelled out in the DNA of the plant. One of the reactions in the chain of chemical processes that comprise photosynthesis must take place in 3 trillionths of a second or the process would fail. Without the food and oxygen that the photosynthesis process produces there would be no life, as we know it, on Earth.

And, finally, most of the carbon dioxide being emitted into the atmosphere is coming from the lakes and oceans—not man. It isn't very likely that any efforts by man will control the amount of carbon dioxide coming from the waters of the Earth.

CHAPTER 4

WEATHER PHENOMENA

The purveyors of the current climate change scare tactics continue to point to the hurricanes, typhoons, wildfires, floods, drought, tornados and other weather phenomena as proof of their views—regardless of the historical facts.

Changes in weather and weather phenomena have always occurred. Almost one hundred years ago the southwest United States was impacted by a very significant event caused by a lack of rain over an extended period, and acerbated by the farming methods in use at the time, which resulted in the "Dust Bowl" that drove a quarter million people from the land. Successive dust storms occurred, culminating in a monster storm on Black Sunday, April 14, 1935, that blew east across the United States dropping tons of dust all the way to the East Coast. It's estimated that the storm carried 300,000,000 tons of dirt, more than twice the amount moved during the construction of the Panama Canal.

Although recent measurements have shown a slight warming trend in the earth's climate this is no different than the warm-cold cycles of past history . The simple fact is that weather events such as droughts, excessive rain periods, hurricanes, tornados and typhoons will always be with us. Ice packs at the north and south poles will decrease and increase as they always have.

Since the significant increase in global warming predicted by climatologists using their models has not occurred they were forced to change their terminology. They have now revised their outcomes to speak in terms of climate change instead of global warming. Using the new terminology they can now attribute all manner of weather phenomena to the increase in carbon levels—without proof! By this simple change in terminology they can attribute cold spells, floods, snow storms, etc. to climate change.

Scare tactics, such as the fallacious claims of decreasing polar bear populations, will continue to be propagated—and proven false. Recent studies have shown that the polar bear population has quadrupled in the past ten years!

The simple fact is that man will continue to experience extreme weather events and climate change in the future—and continue to react to them as they always have—by adapting to it. Just as cave men adapted to the colder and warmer climate periods by moving south or north as required., modern man has adapted by insulating homes and using energy for heating and air conditioning. If the climate warms, Greenland may again experience prolific plant growth and the growing periods in Canada and Russia will extend. Instead of fighting any climate change we must adapt to it.

"Experts know that the worst-case climate projections show minimal impact on the overall economy. Buried in the Intergovernmental Panel on Climate Change 2014 report is a chart showing that a global temperature rise of 5 degrees Fahrenheit would have a global economic impact of about 3% in 2100, negligibly diminishing projected global growth over that period to 385% from 400%. There are many reasons to be concerned about a changing climate, including disparate impact across industries and regions. But national economic catastrophe isn't one of them. It should concern anyone who supports well-informed public and policy discussions that the report's authors, reviewers and media coverage obscured such an important point.—Steven Koonin

The following two chapters will lay out the choice before the American public in the coming years. The Green New Deal is the plan of those who consider our current way of life abhorrent due to the production of carbon dioxide in our daily activities. Following the chapter laying out the Green New Deal will be my response to its provisions.

CHAPTER 5

THE "GREEN NEW DEAL"

Shown below is House Resolution 109, the "Green New Deal". As of early 2021 this resolution has not come up for a vote in the House of Representatives. Frankly, it's doubtful if it ever will since it's ridiculous on its face. However, since the Republicans controlled the Senate in the 116th Congress it was put up for a vote to see if those Senators who said they supported the socialistic thrust of this climate change agenda would vote for it. The final vote was 57-0 <u>against</u> the resolution with 43 Democrats voting "Present". To their credit four Democrats joined all the Republicans voting against it.

A critique of this assault on our common sense follows in Chapter 6.

116TH CONGRESS
1ST SESSION

H. RES. 109

Recognizing the duty of the Federal Government to create a Green New Deal.

IN THE HOUSE OF REPRESENTATIVES

FEBRUARY 7, 2019

Ms. OCASIO-CORTEZ (for herself, Mr. HASTINGS, Ms. TLAIB,
Mr. SERRANO, Mrs. CAROLYN B. MALONEY of New York,
Mr. VARGAS, Mr. ESPAILLAT, Mr. LYNCH, Ms. VELÁZQUEZ,
Mr.BLUMENAUER, Mr. BRENDAN F. BOYLE of Pennsylvania,
Mr. CASTRO of Texas, Ms. CLARKE of New York,
Ms. JAYAPAL, Mr. KHANNA, Mr. TED LIEU of California,
Ms. PRESSLEY, Mr. WELCH, Mr. ENGEL, Mr. NEGUSE,
Mr. NADLER, Mr. MCGOVERN, Mr. POCAN, Mr. TAKANO,
Ms. NORTON, Mr. RASKIN, Mr. CONNOLLY, Mr. LOWENTHAL,
Ms. MATSUI, Mr. THOMPSON of California, Mr. LEVIN of
California, Ms. PINGREE, Mr. QUIGLEY, Mr. HUFFMAN,
Mrs. WATSON COLEMAN, Mr. GARCÍA of Illinois,
Mr. HIGGINS of New York, Ms. HAALAND, Ms. MENG,
Mr. CARBAJAL, Mr. CICILLINE, Mr. COHEN, Ms. CLARK of
Massachusetts, Ms. JUDY CHU of California, Ms. MUCARSEL-
POWELL, Mr. MOULTON, Mr. GRIJALVA, Mr. MEEKS,
Mr. SABLAN, Ms. LEE of California, Ms. BONAMICI, Mr. SEAN
PATRICK MALONEY of New York, Ms. SCHAKOWSKY,
Ms. DELAURO, Mr. LEVIN of Michigan, Ms. MCCOLLUM,
Mr. DESAULNIER, Mr. COURTNEY, Mr.LARSON of Connecticut,
Ms. ESCOBAR, Mr. SCHIFF, Mr. KEATING, Mr. DEFAZIO,
Ms. ESHOO, Mrs. TRAHAN, Mr. GOMEZ, Mr. KENNEDY, and
Ms. WATERS) submitted the following resolution; which was
referred to the Committee on Energy and Commerce, and in
addition to the Committees on Science, Space, and Technology,
Education and Labor, Transportation and Infrastructure,
Agriculture, Natural Resources, Foreign Affairs, Financial
Services, the Judiciary, Ways and Means, and Oversight and
Reform, for a period to be subsequently determined by the

Speaker, in each case for consideration of such provisions as fall within the jurisdiction of the committee concerned

---

RESOLUTION

cognizing the duty of the Federal Government to create a Green New Deal.

Whereas the October 2018 report entitled "Special Report on Global Warming of 1.5 °C" by the Intergovernmental Panel on Climate Change and the November 2018 Fourth National Climate Assessment report found that—

(1) human activity is the dominant cause of observed climate change over the past century;

(2) a changing climate is causing sea levels to rise and an increase in wildfires, severe storms, droughts, and other extreme weather events that threaten human life, healthy communities, and critical infrastructure;

(3) global warming at or above 2 degrees Celsius beyond preindustrialized levels will cause—

(A) mass migration from the regions most affected by climate change;

(B) more than $500,000,000,000 in lost annual economic output in the United States by the year 2100;

(C) wildfires that, by 2050, will annually burn at least twice as much forest area in the western United States than was typically burned by wildfires in the years preceding 2019;

(D) a loss of more than 99 percent of all coral reefs on Earth;

(E) more than 350,000,000 more people to be exposed globally to deadly heat stress by 2050; and

(F) a risk of damage to $1,000,000,000,000 of public infrastructure and coastal real estate in the United States; and

(4) global temperatures must be kept below 1.5 degrees Celsius above preindustrialized levels to avoid the most severe impacts of a changing climate, which will require—

(A) global reductions in greenhouse gas emissions from human sources of 40 to 60 percent from 2010 levels by 2030; and

(B) net-zero global emissions by 2050;

Whereas, because the United States has historically been responsible for a disproportionate amount of greenhouse gas emissions, having emitted 20 percent of global greenhouse gas emissions through 2014, and has a high technological capacity, the United States must take a leading role in reducing emissions through economic transformation;

Whereas the United States is currently experiencing several related crises, with—

(1) life expectancy declining while basic needs, such as clean air, clean water, healthy food, and adequate health care, housing, transportation, and education, are inaccessible to a significant portion of the United States population;

(2) a 4-decade trend of wage stagnation, deindustrialization, and antilabor policies that has led to—

(A) hourly wages overall stagnating since the 1970s despite increased worker productivity;

(B) the third-worst level of socioeconomic mobility in the developed world before the Great Recession;

(C) the erosion of the earning and bargaining power of workers in the United States; and

(D) inadequate resources for public sector workers to confront the challenges of climate change at local, State, and Federal levels; and

(3) the greatest income inequality since the 1920s, with—

(A) the top 1 percent of earners accruing 91 percent of gains in the first few years of economic recovery after the Great Recession;

(B) a large racial wealth divide amounting to a difference of 20 times more wealth between the average white family and the average black family; and

(C) a gender earnings gap that results in women earning approximately 80 percent as much as men, at the median;

Whereas climate change, pollution, and environmental destruction have exacerbated systemic racial, regional, social, environmental, and economic injustices (referred to in this preamble as "systemic injustices") by disproportionately affecting indigenous peoples, communities of color, migrant communities, deindustrialized communities, depopulated rural communities, the poor, low-income workers, women, the elderly, the unhoused, people with disabilities, and youth (referred to in this preamble as "frontline and vulnerable communities");

Whereas, climate change constitutes a direct threat to the national security of the United States—

(1) by impacting the economic, environmental, and social stability of countries and communities around the world; and

(2) by acting as a threat multiplier;

Whereas the Federal Government-led mobilizations during World War II and the New Deal created the greatest middle class that the United States has ever seen, but many members of frontline and vulnerable communities were excluded from many of the economic and societal benefits of those mobilizations; and

Whereas the House of Representatives recognizes that a new national, social, industrial, and economic mobilization on a scale not seen since World War II and the New Deal era is a historic opportunity—

(1) to create millions of good, high-wage jobs in the United States;

(2) to provide unprecedented levels of prosperity and economic security for all people of the United States; and

(3) to counteract systemic injustices: Now, therefore, be it

Resolved, That it is the sense of the House of Representatives that—

(1) it is the duty of the Federal Government to create a Green New Deal—

(A) to achieve net-zero greenhouse gas emissions through a fair and just transition for all communities and workers;

(B) to create millions of good, high-wage jobs and ensure prosperity and economic security for all people of the United States;

(C) to invest in the infrastructure and industry of the United States to sustainably meet the challenges of the 21st century;

(D) to secure for all people of the United States for generations to come—

(i) clean air and water;

(ii) climate and community resiliency;

(iii) healthy food;

(iv) access to nature; and

(v) a sustainable environment; and

(E) to promote justice and equity by stopping current, preventing future, and repairing historic oppression of indigenous peoples, communities of color, migrant communities, deindustrialized communities, depopulated rural communities, the poor, low-income workers, women, the elderly, the unhoused, people with disabilities, and youth (referred to in this resolution as "frontline and vulnerable communities");

(2) the goals described in subparagraphs (A) through (E) of paragraph (1) (referred to in this resolution as the "Green New Deal goals") should be accomplished through a 10-year national mobilization (referred to in this resolution as the "Green New Deal mobilization") that will require the following goals and projects—

(A) building resiliency against climate change-related disasters, such as extreme weather, including by leveraging funding and providing investments for community-defined projects and strategies;

(B) repairing and upgrading the infrastructure in the United States, including—

(i) by eliminating pollution and greenhouse gas emissions as much as technologically feasible;

(ii) by guaranteeing universal access to clean water;

(iii) by reducing the risks posed by climate impacts; and

(iv) by ensuring that any infrastructure bill considered by Congress addresses climate change;

(C) meeting 100 percent of the power demand in the United States through clean, renewable, and zero-emission energy sources, including—

(i) by dramatically expanding and upgrading renewable power sources; and

(ii) by deploying new capacity;

(D) building or upgrading to energy-efficient, distributed, and "smart" power grids, and ensuring affordable access to electricity;

(E) upgrading all existing buildings in the United States and building new buildings to achieve maximum energy efficiency, water efficiency, safety, affordability, comfort, and durability, including through electrification;

(F) spurring massive growth in clean manufacturing in the United States and removing pollution and greenhouse gas emissions from manufacturing and industry as much as is technologically feasible, including by expanding renewable energy manufacturing and investing in existing manufacturing and industry;

(G) working collaboratively with farmers and ranchers in the United States to remove pollution and greenhouse gas emissions from the agricultural sector as much as is technologically feasible, including—

(i) by supporting family farming;

(ii) by investing in sustainable farming and land use practices that increase soil health; and

(iii) by building a more sustainable food system that ensures universal access to healthy food;

(H) overhauling transportation systems in the United States to remove pollution and greenhouse gas emissions from the transportation sector as much as is technologically feasible, including through investment in—

(i) zero-emission vehicle infrastructure and manufacturing;

(ii) clean, affordable, and accessible public transit; and

(iii) high-speed rail;

(I) mitigating and managing the long-term adverse health, economic, and other effects of pollution and climate change, including by providing funding for community-defined projects and strategies;

(J) removing greenhouse gases from the atmosphere and reducing pollution by restoring natural ecosystems through proven low-tech solutions that increase soil carbon storage, such as land preservation and afforestation;

(K) restoring and protecting threatened, endangered, and fragile ecosystems through locally appropriate and science-based projects that enhance biodiversity and support climate resiliency;

(L) cleaning up existing hazardous waste and abandoned sites, ensuring economic development and sustainability on those sites;

(M) identifying other emission and pollution sources and creating solutions to remove them; and

(N) promoting the international exchange of technology, expertise, products, funding, and services, with the aim of making the United States the international leader on climate action, and to help other countries achieve a Green New Deal;

(3) a Green New Deal must be developed through transparent and inclusive consultation, collaboration, and partnership with frontline and vulnerable communities, labor unions, worker cooperatives, civil society groups, academia, and businesses; and

(4) to achieve the Green New Deal goals and mobilization, a Green New Deal will require the following goals and projects—

(A) providing and leveraging, in a way that ensures that the public receives appropriate ownership stakes and returns on investment, adequate capital (including through community grants, public banks, and other public financing), technical expertise, supporting policies, and other forms of assistance to communities, organizations, Federal, State, and local government agencies, and businesses working on the Green New Deal mobilization;

(B) ensuring that the Federal Government takes into account the complete environmental and social costs and impacts of emissions through—

(i) existing laws;

(ii) new policies and programs; and

(iii) ensuring that frontline and vulnerable communities shall not be adversely affected;

(C) providing resources, training, and high-quality education, including higher education, to all people of the United States, with a focus on frontline and vulnerable communities, so that all people of the United States may be full and equal participants in the Green New Deal mobilization;

(D) making public investments in the research and development of new clean and renewable energy technologies and industries;

(E) directing investments to spur economic development, deepen and diversify industry and business in local and regional economies, and build wealth and community ownership, while prioritizing high-quality job creation and economic, social, and environmental benefits in frontline and vulnerable communities, and deindustrialized communities, that may otherwise struggle with the transition away from greenhouse gas intensive industries;

(F) ensuring the use of democratic and participatory processes that are inclusive of and led by frontline and vulnerable communities and workers to plan, implement, and administer the Green New Deal mobilization at the local level;

(G) ensuring that the Green New Deal mobilization creates high-quality union jobs that pay prevailing wages, hires local workers, offers training and advancement opportunities, and guarantees wage and benefit parity for workers affected by the transition;

(H) guaranteeing a job with a family-sustaining wage, adequate family and medical leave, paid vacations, and retirement security to all people of the United States;

(I) strengthening and protecting the right of all workers to organize, unionize, and collectively bargain free of coercion, intimidation, and harassment;

(J) strengthening and enforcing labor, workplace health and safety, antidiscrimination, and wage and hour standards across all employers, industries, and sectors;

(K) enacting and enforcing trade rules, procurement standards, and border adjustments with strong labor and environmental protections—

(i) to stop the transfer of jobs and pollution overseas; and

(ii) to grow domestic manufacturing in the United States;

(L) ensuring that public lands, waters, and oceans are protected and that eminent domain is not abused;

(M) obtaining the free, prior, and informed consent of indigenous peoples for all decisions that affect indigenous peoples and their traditional territories, honoring all treaties and agreements with indigenous peoples, and protecting and enforcing the sovereignty and land rights of indigenous peoples;

(N) ensuring a commercial environment where every businessperson is free from unfair competition and domination by domestic or international monopolies; and

(O) providing all people of the United States with—

(i) high-quality health care;

(ii) affordable, safe, and adequate housing;

(iii) economic security; and

(iv) clean water, clean air, healthy and affordable food, and access to nature.

CHAPTER 6

CRITIQUE OF THE SOCIALISTIC WISH LIST

It's difficult to know where to begin in the critique of the piece of proposed legislation labeled the Green New Deal.

Before addressing each of its proposals it might be useful to first start with the subject of its implementation cost—estimated at approximately $94 trillion. The supporters of the plan are calling for a vast increase in taxes as a starting point. To quote the late Prime Minister of Britain, Margaret Thatcher, "The problem with socialism is that eventually you run out of other people's money."

Since the implementation of its provisions will destroy our capitalistic economy it's difficult to determine where the additional monies to pay for the completion of the socialistic paradise will come from since it doesn't appear there are any provisions in the legislation for the planting of groves of money trees. If their intention is to start running the Treasury's money presses around the clock our streets we'll quickly begin looking like Venezuela where it takes a wheelbarrow of money to buy a loaf of bread. For the majority of our citizens life will quickly revert to the standard of the Middle Ages where life was "nasty, brutish and short".

Ask the Venezuelans how things turned out for them when they converted from a capitalistic to a socialistic economy.

The opening pages of the House Resolution begins with a series of lies:

- Human activity has <u>not</u> been proven to be the dominant cause of observed climate change in the past century. Where's the proof for that outrageous statement?

- Although there may be a small increase in sea levels where is the proof that this has led to an increase in wildfires and extreme weather events?

- Where is the proof that a couple of degrees of temperature change will cause the dire effects they prophesize? The $500 trillion figure they attribute as the economic loss in the United States is pure fantasy. They cannot back up that figure. Even the UN Panel disagrees with them. See the quotation at the end of Chapter 4.

- The other dire figures they state in their opening statements are unable to be supported by scientists and are pulled from thin air.

Having opened their resolution with a plethora of lies they then propose their solutions:

- A global reduction of 40-60% in carbon dioxide by 2030.

- Net-zero global emissions by 2050

Not only are these reductions impossible for the United States, the resolution specifies these are <u>global</u> goals. Good luck imposing these reductions on the Chinese, Russians, Iranians, North Koreans and the rest of the world. I'm sure the other countries of the world won't spend much time dealing with the demands of our climate change zealots. Unfortunately, we have to.

The resolution then goes on to list the problems we currently have in the United States, none of which have anything to do with global warming. The Green New Deal is being put forward to correct all perceived problems in our capitalistic society and convert it to socialism, using identity politics as the driving force—rich against poor, black against white, men against women, old against young, the list is endless. It's also obvious that the proponents of this legislation are living in an alternate universe. We do have problems of homelessness, income inequality, medical coverage and areas of poverty that must be addressed, but the United States is still the "shining beacon on the hill" that many in the world want to come to . We provide the highest standard of living and more freedom for our citizens than any other major country on Earth.

The resolution then goes on to list all the objectives that it wishes to fulfill including clean air and water, healthy food, high wage jobs, access to nature . . . All that appears to be left out is "mom, baseball and apple pie".

What follows are the specific actions the resolution envisages including:

- Attaining 100% emission-free power generation, which means all coal, oil and gas power generating plants would be deactivated. Since nuclear energy is not presented as an acceptable choice that pretty much leaves solar and wind as the major sources. All leading experts in power generation say this objective is absolutely impossible for many reasons, including the lack of sufficient battery resources to store energy when the wind isn't blowing and the sun isn't shining. No known electrical network grid is available to handle a 100% solar/wind array of power generation facilities.

- The upgrading of all presently constructed buildings in the United States to make them energy efficient. We can say goodbye to glass and steel structures that currently comprise the bulk of our major city business districts. Again, we have to ask—where is all the money coming from to retrofit all of these buildings?

- Removing all pollution and carbon dioxide generation from manufacturing plants in the

United States. There goes our balance of trade since apparently we will now be importing all manufactured goods from countries who aren't as stupid as we are.

- Removing pollution and greenhouse gas emissions from the agricultural sector. So much for you steak eaters since those farting cows will have to go. I'm not sure any meat will remain on future menus since all farm animals generate pollution.

- Overhauling the transportation systems in the United States. All internal combustion cars will be forbidden. The Indianapolis 500 and stock car races will definitely be attractions of the past. It appears long-haul trucking will have to go, since high-speed rail will be replacing trucks and planes. Good luck with that. In addition to the massive strike actions by truck drivers which will certainly occur, the building of high-speed rail will be the biggest boondoggle of them all if the example set by one construction attempt is indicative—that of the proposed high-speed rail project between Los Angeles and San Francisco.

"I would suggest that the projections might be a bit optimistic."—Karl Compton, Letters to the Editor in The Wall Street Journal, who pointed out that the plans for the construction of a $40 billion dollar bullet train system from Los Angeles to San Francisco projected 41 million riders per year (40% of

breakeven). "... that works out to 112,328 riders per day or 4680 per hour. With 150 passengers packed cheek-to-jowl into each car, that's a 31-car train leaving every hour of the day and night."

Since a round-trip air ticket currently goes for approximately $200, one has to question how many travelers would want to pony up the $400 for a one-way train ticket. Anyone else have a multibillion-dollar boondoggle idea to build a terrorist target in an earthquake prone region? Note: In early 2019 California Governor Newsom canceled the partially completed bullet train to Los Angeles when its projected cost rose to $77 billion and its completion date was 100 years in the future at its current construction rate.

Since it appears plane travel is also on the chopping block we will have to come up with a means of travel to Hawaii, Asia and Europe. Underwater trains, anyone?

Economic security from cradle to grave is guaranteed with no requirement that you actually need to work. One again has to wonder where the monies will come from to fulfill this guarantee . If no one has to work, there won't be anyone to collect taxes from. It's all a little confusing—harebrained or half-baked would be a better description. At this point it calls to mind a humorous quotation regarding socialism:

"The trouble with socialism is that it would take too many evenings."—Oscar Wilde, alluding to the fact that the workers would have to decide what to make and what to charge, decisions currently made by the marketplace in a capitalistic society.

CHAPTER 7

THE DEVIL'S ADVOCATE

At this point it would be worthwhile to lay out the position of the Devil's Advocate. In the Roman Catholic Church a Devil's Advocate is assigned to present evidence against a case for the pronouncement of sainthood for a person.

The major argument in favor of following the path laid out by the global warming proponents is this: What if man's activities are, in fact, significantly contributing to global warming. Wouldn't it be better to be safe than sorry, and take steps to eliminate carbon dioxide?

It appears the answer to this question is "yes". However, the climatologists have come to the conclusion that even if the United States takes drastic steps, starting today, these actions will have no impact on the global warming they are projecting. Therefore, it makes much more sense to adapt to the changing climate and not try to stop it.

CHAPTER 8

THE FORK IN THE ROAD

We are facing the proverbial "fork in the road". One path leads to a drastic restructuring of our lives and vast controls by government over our movements and beliefs—in short the dystopian socialistic anthill environment portrayed by George Orwell in his novel *1984*. Big Brother will be watching to insure we conform to the party line of the Green New Deal. This is the path laid out for us by a 41-year old ex-bartender, and an ex-Vice President who received a "D" in his Natural Sciences college course. Their road map will completely destroy the American economy by imposing a socialistic government on us.

The other path is the one that all makes the most sense—a path that realizes that if global warming is occurring we must adapt to it Money should not be spent attempting to stop a force that is not stoppable. We would do much better applying funds to the actual social problems we do face, and to medical research on cancer, Alzheimer's and other diseases.

To follow the plan outlined in the Green New Deal is simply insane. It makes no sense to wreck the American economy by taking steps that will have no discernible effect on the earth's temperature. We must call out this "Deal" for what it simply is—a pathway to socialism in the United States.

It seems fitting to end this chapter with a quote from a brilliant commentator who left this earth far too early:

"No matter how far the ideological pendulum swings in the short term, in the end, the bedrock common sense of the American people will prevail."—Charles Krauthammer

EPILOGUE

Although this book was concerned with debunking the claim of man-made global warming, there is, in fact, a potential catastrophic event we should prepare for—an extinction event arising from an asteroid impact on earth. similar to the one that wiped out the dinosaurs, 65 million years ago.   If we are able to identify an asteroid headed for Earth in enough time we could take action to affect its trajectory—if we are prepared! Although there are   some ongoing efforts by astronomers to identify possible dangerous asteroids, a much larger comprehensive effort undertaken jointly by all nations should be initiated to identify and intercept a potential killer asteroid.

A very rare event occurred on February 15, 2013, when a large 10,000 ton meteor traveling at 43,000 miles per hour, exploded over Chelyabinsk, Russia, injuring 1500 people. Miraculously, no one was killed.   Coincidentally, on that same day, an asteroid estimated at 40,000 tons, passed within 17,200 miles of the Earth's surface, closer than many of the satellites above the Earth—also a very rare event. Had it struck the Earth it would have caused catastrophic damage. The probability of both these events happening on the same day is astronomical (no pun intended). Perhaps someone was giving us a message.

RELEVANT QUOTES

"The whole aim of practical politics is to keep the populace alarmed (and clamorous to be led to safety} by menacing it with an endless series of hobgoblins, all of them imaginary"—H. L. Mencken.   Man-made climate change is the latest hobgoblin foisted on the American public.

"Global warming is largely a natural phenomenon. The world is wasting stupendous amounts of money on trying to fix something that can't be fixed."—Dr. David Bellamy, Botanist and Environmentalist

" . . . the innate behavior of the climate system imposes limits on the ability to predict its evolution."—From the latest report of the Inter-governmental Panel on Climate Change, published in March 2014. This is an astounding statement by the IPCC, given its years of "crying wolf" regarding the impending doom of the planet due to man-made global warming, contrary to the empirical evidence showing no significant global warming in the last twenty years.

"When we make (mathematical) models involving human beings we are trying to force the ugly stepsister's foot into Cinderella's pretty glass slipper. It doesn't fit without cutting out some of the essential parts."—Emanuel Derman, from his book, *Models Behaving Badly*. This analogy also applies to the climate change models in use today.

"I used to agree with these dramatic warnings of climate disaster . . . However, a few years ago I decided to look more closely at the science and it astonished me. In fact, there is no evidence of humans being the cause. There is, however, overwhelming evidence of natural causes such as changes in the output of the sun."—Dr. Ian D. Clark, Paleoclimatologist

"The causes of these global changes is fundamentally due to the Sun and its effect on the Earth as it moves about in its orbit—not from man-made activities."—Dr. William W. Vaughn, Award Winning NASA Atmospheric Scientist

"Satellites have recorded a roughly 14% increase in greenery on the planet over the past 30 years, in all types of ecosystems, partly as a result of man-made CO2 emissions, which enable plants to grow faster and use less water. . . Almost every global environmental scare of the past half-century proved exaggerated, including the population "bomb", pesticides, acid rain, the ozone hole, falling sperm counts, genetically engineered crops and killer bees. In every case, institutional scientists gained a lot of funding from the scare and then quietly converged on the view that the problem was much more moderate than the extreme voices had argued. Global warming is no different."—Matt Ridley

"How can you say to the hungry of this earth—how can you say to those who don't enjoy the luxury that we all do and that the developed world in general does, how can you tell those folks, 'Sorry about your luck.' You know this is an indulgence of the rich and it is not just scientifically indefensible, it is morally indefensible."— From a speech by Purdue University President Mitch Daniels, Jr., attacking the anti-GMO movement in its attempts through junk science and false claims to stifle new technologies which would allow food abundance for all. The same statement is applicable to those who would deny oil, gas and coal to provide economic power to these underdeveloped economies, where a significant part of their populace is without electricity.

"Soviet life was a pail of milk of human kindness, with a dead rat at the bottom."—Russian writer Vladimir Nabokov

"Scratch an intellectual, and you will find a would-be aristocrat who loathes the sight, the sound and the smell of common folk."—Eric Hoffer, from his book *The True Believer*

"If we extend unlimited tolerance even to those who are intolerant, if we are not prepared to defend a tolerant society against the onslaught of the intolerant, then the tolerant will be destroyed and tolerance with them."—Karl Popper, explaining "the paradox of tolerance"

"Banning DDT killed more people than Hitler."—Author Michael Crichton, alluding to the fact that the DDT ban allowed mosquitoes to infect the malaria virus into millions of African children. Another calamitous example of the rule of unintended consequences of precipitous actions taken to solve current problems.

"I firmly believed that the ends justified the means. Our great goal was the universal triumph of Communism, and for the sake of the goal everything was permissible—to lie, to steal, to destroy hundreds of thousands and even millions of people, all those that were hindering our work, or could hinder it, everyone who stood in our way. And to hesitate and doubt about all this was to give in to 'intellectual squeamishness' and 'stupid liberalism'."—Said by a Russian Communist, following the 1917 revolution

"Some might say the USSR and Venezuela never implemented true socialism. Even so, that merely confirms another argument against socialism: it has to rely on imperfect, self-interested people to staff its bureaucracies, plan the economy, and oversee an excessively centralized administration. Dreams of liberation and progress quickly deteriorate into tyranny and dictatorship. Once fallible men get a taste of power, they are reluctant to relinquish it."—Jeff Cimino

"Underlying most arguments against the free market is a lack of belief in freedom itself."—Milton Friedman

"Since heat-related deaths are generally much fewer than cold-related deaths, the overall effect of global warming on health can be expected to be a beneficial one."—From a 2004 study by William Keating and Gavin Donaldson

"The first lesson of economics is scarcity: There is never enough of anything to satisfy all those who want it. The first lesson of politics is to ignore the first lesson of economics."—Thomas Sowell.

"A civilization that feels guilty for everything it is and does will lack the energy and conviction to defend itself."—French philosopher Jean Francois Revel. Quoted by United Nations Ambassador Jeanne Kirkpatrick, a lifelong Democrat, as she electrified the 1984 Republican National Convention, when she described the Democratic National Convention, held some weeks earlier, as a meeting of the "Blame America First" crowd. This "crowd" has now returned with a vengeance in the 21st century, blaming America for much of the global warming they say is occurring.

"(Socialism that is democratic) has never existed, and in its absence socialism without democracy becomes the shimmering goal (of many liberals), because socialism cannot be given up. Therein lies the final irony. Just as democratic socialism is an illusion, so too is socialism itself the pipe dream that never dies. It promises harmony and abundance; it has always instead produced strife and penury. Little wonder, then, that free people have never chosen it."—Joshua Muravchik, in a commentary in *The Weekly Standard.*

"No responsible leader can put the workers—and the people—of their country at this debilitating and tremendous disadvantage."—President Donald Trump, speaking of the costs of the Paris climate agreement, as he withdrew the United States from the agreement.

"There is no Plan B because there is no Planet B."—French President Emmanuel Macron lecturing President Donald Trump, after Trump had announced that the United States was withdrawing from the Paris climate agreement. Within months Macron had backtracked from his plan to impose new carbon taxes on the French people as part of his fulfillment of the climate agreement's requirements, in the face of massive riots throughout France. Apparently there will have to be a Plan B.

"Socialist economic proposals are recipes for economic stagnation. If the state owns corporations, there is no competition, only rivalries among people with political power."—Stephen Miller

"My reading of history convinces me that most bad government results from too much government."—President Thomas Jefferson

"I regard consensus science as an extremely pernicious development that ought to be stopped cold in its tracks. Historically, the claim of consensus has been the first refuge of scoundrels; it is a way to avoid debate by claiming that the matter is already settled. Whenever you hear the consensus of scientists agrees on something or other, reach for your wallet, because you're being had. . . . There is no such thing as consensus science. If it's consensus, it isn't science. . . .Nobody believes a weather prediction twelve hours ahead. Now we're asked to believe a prediction that goes out 100 years into the future? And make financial investments based on that prediction? Has everybody lost their minds?"—Author Michael Crichton

"All great movements start as a cause, evolve into a business and end up a racket."—Eric Hoffer, author of *The True Believer*

"Here's the truth, brothers and sisters, there's plenty of money in the world. There's plenty of money in this city. It's just in the wrong hands."—New York City Mayor Bill de Blasio, making a statement typical of a socialist redistributionist who has zero understanding of economics. How many of those who earned the money are going to stick around waiting for him to confiscate it?

"All active mass movements strive to interpose a fact-proof screen between the faithful and the realities of the world. They do this by claiming that the ultimate and absolute truth is already embodied in their doctrine and that there is no truth nor certitude outside it. To rely on the evidence of the senses and reason is heresy and treason."—Eric Hoffer *The True Believer*

I don't aspire to measure the global temperature, nor to estimate the importance of factors which make it. This is not the area of my comparative advantage. But to argue, as it's done by many contemporary environmentalists, that these questions (regarding the negative impact of humans on the environment and our ability to affect it) have already been answered with an unqualified 'yes' and that there is unchallenged scientific consensus about this is unjustified. It is also morally and intellectually deceptive."—Czech President Vaclav Klaus

"The inherent vice of capitalism is the unequal sharing of blessings. Socialism is a philosophy of failure, the creed of ignorance and the gospel of envy; its inherent virtue is the equal sharing of miseries."—Winston Churchill

"You're going to find signs on manufacturing doors, if this bill passes, that say, 'Moved—gone to China'."—Senator Charles Grassley, commenting on the cap-and-trade bill, coming before the Senate, designed to curb greenhouse-gas emissions, by imposing significant costs on manufacturing industries using oil and coal.

"We now know that the extra carbon dioxide and global warmth, no matter what their cause, are resulting in a gradual greening of the Earth. There is some evidence that there has been a slight poleward shift in the habitats of some warm weather species, from the tropics where there is a great diversity of life, to higher latitudes where many of these forms of life could not otherwise survive. Global warming has made weather less severe, and cold weather is known to cause more deaths than hot weather. So why is global warming necessarily a bad thing?"—Roy W. Spencer, from his book *Climate Confusion*.

"If we were directed from Washington when to sow and when to reap, we would soon want bread."—Thomas Jefferson

"The link between the burning of fossil fuels and global warming is a myth. It is time the world's leaders, their scientific advisers and many environmental pressure groups woke up to the fact"

* * *

"...carbon dioxide is *not* the dreaded killer greenhouse gas...It is, in fact, the most important airborne fertilizer in the world... (vital for photosynthesis).

* * *

"The real truth is that the main greenhouse gas—the one that has the most direct effect on land temperatures—is water vapour, 99% of which is entirely natural."—Excerpts from a full page article titled "Global Warming? What a Load of Poppycock!" in the London newspaper, *The Daily Mail*, by Professor David Bellamy, an ardent environmentalist.

"Workers of the world, forgive me."—Graffiti written on the bust of Karl Marx, founder of Communism, in Bucharest, Romania, in a play on the words of the Communist Manifesto, "Workers of the World, Unite!"

"We're going to need oil and gas and coal for a long time if America wants to keep the lights on. What I see are people who want affordable energy. They want strong environmental standards—they want a lot of things—but first and foremost they want affordable energy. And if you want affordable energy, you want oil, gas and coal."—John Watson, CEO, Chevron Oil.

"Actually what makes skeptics (of global warming) skeptical is the accumulating evidence that theories predicting catastrophe from man-made climate change are impervious to evidence."—George Will, 2009, commenting on the trend reported in the *New York Times* that, "...global temperatures have been relatively constant for a decade and may even drop in the next few years."

"The sudden abundance of low-cost natural gas is a gift."—Tim Wirth, president of the United Nations Foundation. With the introduction of a new drilling technique (hydraulic fracturing) developed in the late 1990s, immense deposits of shale gas became recoverable in the United States. The quantity of recoverable natural gas in the United States and Canada is now estimated to last 100 years—and the United States is projected to be an exporter of liquefied natural gas rather than an importer.

"Atlas Shrugged."—Title of Ayn Rand's bestseller that praised unrepentant capitalism, published in 1957. The novel's title refers to the action of the mythological Atlas who carried the world on his shoulders until the weight became too much. The book's underlying thesis was that self-interest is paramount to all other interests and that when a society restricts and taxes its most productive and creative citizens too much in the interest of wealth redistribution, these citizens will simply "throw in the towel", walk away and let the society perish. The book sold 400,000 copies in 2009, double the total of previous sales in any of the previous fifty-two years since its publication.

"When you see that in order to produce you must obtain permission from men who produce nothing . . . and your laws don't protect you against them, but protect them against you . . . you may know that your society is doomed."—From *Atlas Shrugged* by Ayn Rand.

"Communism is the time that countries waste between capitalism and capitalism."—Cuban Carlos Alberto Montaner, commenting on the 52nd year that the Cuban government has wasted waiting for prosperity.

"The capitalistic economy goes out of its way to put ourselves first. (It derives from an egotism) rooted in the 'old brain' which was bequeathed to us by the reptiles that struggled out of the primal slime 500 million years ago."—Karen Armstrong, author of *Twelve Steps to a Compassionate Life.*

". . . those nasty old capitalists, with their vigor, risk-taking, animal spirits and reptilian brains, have created so much wealth for so many societies over so many centuries—and have raised the standard of living for so many people who would otherwise live in grinding poverty—that their efforts, easily considered merely selfish, begin to look downright compassionate."—Eric Felton, in his critical review of the book cited in the previous quote above, in *The Wall Street Journal.*

"Are we never to learn that socialism has its roots in envy and in nothing else."—Norman Douglas

"The most astounding fact about reformers, driven by the purest of motives and most spotless goodwill, is that it does not dawn on them that their programs can make things worse."—Leo Rosten

". . . China and India, . . . have told Obama officials they have no intention of signing on to the rich world's growth-killing obsessions."—*Wall Street Journal* editorial response to the projected U.S. cap-and-tax bill on carbon emissions expected to drive up the cost of U.S. manufacturers and ultimately to result in the movement of the manufacturers overseas. A group of U. S. Senators from states with significant manufacturing employment indicated they wanted a tariff on imports from countries without carbon emissions restrictions, as the price for their support of the bill—an action that would result in a trade war.

"Congress . . . will spurn calls to send billions in 'climate reparations' to China and other countries. Representatives of those nations, when they did not have their hands out in Copenhagen, grasping for America's wealth, clapped their hands for Hugo Chavez and other kleptocrats who denounced capitalism while clamoring for its fruits."—George Will, in a commentary on the Copenhagen Climate Conference.

"Global warming is necessary to prevent a new Ice Age."—Conclusion of an article in *Science* magazine (January 2010).

"We can't wait to hear Mr. Obama tell Americans that he wants them to pay higher taxes so the U. S. can pay China to become more energy efficient and thus more economically competitive."—From a *Wall Street Journal* editorial commenting on the announcement of an initiative by the Obama administration at a climate conference in Denmark to help raise an annual $100 billion fund to assist developing nations, including China, to become more energy efficient and thus help reduce "man-made" global warming, an effort most climatologists agree would have little or no impact on the earth's climate.

"The frog is coming awake at just the last moment. He is jumping out of the water."—Peggy Noonan, comparing the American people, on the issue of excessive government spending, to a frog in a pot of water—it was predicted that the rising heat would lull the frog, and when the water came full boil, it wouldn't be able to jump out. This "jumping frog" analogy can just as well be applicable to the Green New Deal.

"Occupy Everything, Death to Capitalism." —Large black banner carried by protesters in Oakland, California. My question is "Why are they here?" They should all be applying for visas to go to North Korea, the anti-capitalist paradise of their dreams.

"Democrats believe in the welfare state before they believe in capitalism. The assumption is that there is some kind of perpetual engine of economic prosperity in America that is just going to continue. And therefore they are able to take from those who create and give it to those who don't."—Republican Eric Cantor, House Majority Leader

"So that the record of history is absolutely crystal clear. That there is no alternative way, so far discovered, of improving the lot of the ordinary people that can hold a candle to the productive activities that are unleashed by a free enterprise system."—Milton Friedman

"Somehow we have got to figure out how to boost the price of gasoline to the levels in Europe."—Steven Chu, U. S. Secretary of Energy, under the Obama administration.

"Everyone must be the same and have the same. Social justice means we deny ourselves many things so that others may have to do without them as well."—Sigmund Freud, commenting on class envy.

"I have been a member of the European Parliament for twelve years. I am living in your future or at least the future your present leaders seem intent on taking you. Believe me, my friends, you are not going to enjoy it. . . . We are at the end of the road you have just set out along. . . . We're screeching towards the cliff. You know what? We look up and what do we see in our rear view mirror? We see you trying to overtake us, accelerating frantically in the direction we have been going. My friends, there is still time to turn aside."—Daniel Hannon, one of Great Britain's representatives to the European Parliament, and one of the few dissenting members to its current policies, in a 2012 speech to CPAC (Conservative Policy Action Council) in Washington, D. C.

"Socialists can provide you shelter, fill your belly with bacon and beans, treat you when you're ill—all the things guaranteed to a prisoner or a slave."—Ronald Reagan

" . . . the greatest and most successful pseudoscientific fraud I have seen in my long life as a physicist."—Harold Lewis, University of California Emeritus Professor of Physics, castigating the science behind global warming. Ivar Giaever, 1976 Nobel Laureate, supporting Lewis, stated that a .8 degree change, on the Kelvin scale from -288.0 to -288.8 in 150 years, meant to him that the earth's temperature was amazingly stable.

"Environmental regulations are seen to be the number one risk to reliability over the next one to five years."—From the report of the North American Electric Reliability Corporation (NERC), an independent advisory body, appointed by Congress, to monitor the reliability of the nation's electrical grid. Stung by the criticism, Obama administration EPA officials initiated an audit of the NERC, in a "kill the messenger" reaction.

"As Americans we must always remember we all have a common enemy, an enemy who is dangerous, powerful and relentless. I refer, of course, to the federal government."—Dave Barry. Marquette University demanded that one of its graduate students remove this quotation, which had been posted on the student's door.

"Can you imagine 400 million people who do not have a light bulb in their home? You cannot, in a democracy, ignore some of these realities and as it happens with the resources of coal that India has, we really don't have any choice but to use coal."—Rajandra Pachauri, Indian academic, chairman of the United Nations Intergovernmental Panel On Climate Change, facing the reality that coal will not be displaced as one of the primary sources of electrical generation anytime in the foreseeable future.

"Socialism is like a dream. Sooner or later you wake up to reality."—Winston Churchill

"If you think health care is expensive now, wait until you see what it costs when it's free!"—P. J. O'Rourke

"You are never dedicated to something you have complete confidence in. (No one is fanatically shouting that the sun is going to rise tomorrow. They know it's going to rise tomorrow.) When people are fanatically devoted to political or religious faiths or any other kinds of dogma or goals, it's always because these dogmas or goals are in doubt."—Robert M. Pirsig

"I wish that during the years that I was in public office, I had had this firsthand experience. We intuitively know that to create job opportunities, we need entrepreneurs who will risk their capital against an expected payoff. Too often, however, public policy does not consider whether we are choking off those opportunities."— Former Democratic presidential candidate George McGovern, who had recently lost his entire investment in a Connecticut motor inn, which he attributed to burdensome governmental regulations.

"One has to belong to the intelligentsia to believe things like that: no ordinary man could be such a fool."—George Orwell, commenting on some of the theories of contemporary scientists. Quotes such as this are timeless.

"The pause in the rise of the global average temperature may have already lasted 17 years, depending on which data set you look at."—Rajendra Pachauri, chairman of the Intergovernmental Panel on Climate Change. Based on hindsight, some climate change models have overestimated warming by 100% over the past 20 years.

"It comes from outer space, lands in the woods, and it's the size of an orange or cantaloupe. By the end of the movie, what happens is it's enveloping diners and houses. As the blob rolls along eating folks, it got bigger and bigger and bigger. And it never got smaller. It consumed more and more. And that's how I kind of see the federal government. Instead of eating people and diners, it's eating our liberty. I'm not against government. I'm against this ever-expanding government that doesn't know its limits."—South Carolina attorney general Alan Wilson, using the symbolism of the movie *The Blob*, to describe the ever encroaching reach of the federal government.

"I think we have to stop considering *Climate Research* as a legitimate peer-review journal . . . Perhaps we should encourage our colleagues in the climate research community to no longer submit to, or cite papers in, this journal."—Climate researcher Professor Michael Mann to a colleague in a leaked e-mail, commenting on the publication of an article questioning global warming theories in the cited journal. Professor Mann was the creator of the infamous "hockey stick" graph showing a sudden upward turn in world average temperature since the Industrial Revolution—a depiction that was later shown to be based on erroneous statistics. This bogus graph was used prominently in former Vice President Gore's award winning documentary "An Inconvenient Truth."

"The (environmental) prophet is not a great soul who admonishes us but a petty fellow who wishes us many misfortunes if we have the gall not to listen to him. Catastrophe is not something that haunts him but his source of joy."—Pascal Bruckner, from his book *The Fanaticism of the Apocalypse.*

"All the foolishness of Bolshevism, Maoism, and Trotskyism are somehow reformulated exponentially in the name of saving the planet."—Pascal Bruckner, from his book *The Fanaticism of the Apocalypse*, descrying the environmentalists of today.

"Just coal by another name."—Environmental activist Mike Tidwell, describing natural gas, ignoring the fact that the vast quantities of natural gas released by the "fracking" method in the United States has resulted in an 11% decrease in carbon dioxide emissions between 2005 and 2011 due to the replacement of coal in power plants. Beside ignoring this fact Tidewell and other activists called for a substantial and costly increase in renewable sources such as solar and wind. Germany, which poured billions into renewable sources has seen its electrical costs soar, and has now turned to coal as its primary source of electricity production.

"A society that puts equality—in the sense of equality of outcome—ahead of freedom will end up with neither equality nor freedom. The use of force to achieve equality will destroy freedom, and the force, introduced for good purposes, will end up in the hands of people who use it to promote their own interests. . . Freedom means diversity but also mobility. It preserves the opportunity for today's disadvantaged to become tomorrow's privileged and, in the process, enables almost everyone, from top to bottom, to enjoy a fuller and richer life."—Milton and Rose Friedman from *Free To Choose*.

"Instead of fostering a system that enables people to help themselves, America is now saddled with a system that destroys value, raises costs, hinders innovation and relegates millions of citizens to a life of poverty, dependency and hopelessness. This is what happens when elected officials believe that people's lives are better run by politicians and regulators than by the people themselves. Those in power fail to see that more government means less liberty, and liberty is the essence of what it means to be American. Love of liberty is the American ideal."—Charles O. Koch, chairman and CEO of Koch Industries, in a criticism written during the Obama administration.

"Observe which side resorts to the most vociferous name-calling and you are likely to have identified the side with the weaker argument and they know it."—Charles R. Anderson, Research Physicist

"You had designers who were constrained and occupied with only one goal, and that was weight and miles per gallon."—Sam Kazman, general counsel of the Competitive Enterprise Council, strongly criticizing the CAFE standards that require higher "miles per gallon" on manufactured cars. According to a 2007 Insurance Institute for Highway Safety 250-500 deaths are attributable to downsized cars attempting to meet more stringent CAFE standards.

"It is not from the benevolence of the butcher, the brewer, or the baker, that we expect our dinner, but from their regard to their own self-interest."—Adam Smith, eighteenth century economist, and author of *The Wealth of Nations*, who theorized that an "invisible hand" underpinned the actions of participants in a market economy.

"(Environmentalism has) become a religion, and religions don't worry much about facts."—James Lovelock, environmental scientist

"In one of the most expensive ironies of history, the expenditure of more than $50 billion on research into global warming has failed to demonstrate any human-caused climate trend, let alone a dangerous one."—Robert Carter, Paleoclimate Scientist"

"Eat less!"—Dame Vivienne Westwood, millionaire British fashion designer, and advocate against genetically modified food, when asked what poor people should do if they couldn't afford to buy organic food. Her response is indicative of the attitude of rich, effete snobs who look down on people who can't afford the luxury of subscribing to their elitist views.

When Milton (Friedman) was starting out, people really believed a state run economy was the most efficient way of promoting growth. Today nobody believes that, except maybe in North Korea. You go to China, India, Brazil, Argentina, Mexico, even Western Europe. Most of the economists under 50 have a free market orientation. Now, there are differences of emphasis and opinion among them. But they're oriented toward the markets. That's a very, very important victory. Will this victory have an effect on policy? Yes, it already has. And in years to come, I believe it will have an even greater impact."—Gary S. Becker, University of Chicago economist

"The environmentalists are for any energy source—unless it actually works."—Stephen Hayward, American Enterprise Institute

"So that the record of history is absolutely crystal clear. That there is no alternative way, so far discovered, of improving the lot of the ordinary people that can hold a candle to the productive activities that are unleashed by a free enterprise system."—Milton Friedman

"For only when our arms are sufficient beyond doubt can we be certain beyond doubt that they will never be employed."—President John F. Kennedy, from his 1960 inaugural speech. Some words to ponder by those who call for the reduction of the America's military forces—especially those involving expenditures designed to keep our military the most modern and unbeatable in the world. Again we should remember that all-too-true quote "Weakness is a provocation." Implementation of the Green New Deal, which advocates extensive military cuts, will certainly result in a much weaker American nation and the military forces that protect it.

"Americans want from government not flights of fancy but sobriety; not ecstatic evocations of dreamlike tomorrows but a tolerably functioning today; not fantasies about a world without scarcities and therefore without choices among our desires and appetites but a mature understanding of the limits to government's proper scope and actual competence."—George Will

"Your Final, Final Warning. This Time We're Serious."—Rapture News, a religious news medium, forecasting the "End of Days." Echoes of these "final, final" warnings by religious zealots have been taken up by the climate change zealots of today.

"The Democrats' Green New Deal calls for a fully renewable electrical power grid. Regardless of the economic or political challenges of bringing this about, it is likely technologically impossible. An electric power grid requires second-by-second balancing between generated supply and consumer demand... This doesn't work for wind and solar because you can't spontaneously increase wind or sunshine.... Fossil fuel turbines, by contrast, naturally compensate for sudden supply outages . . . An all renewables power grid is destined to collapse."—Robert Blohm, member of the North American Electric Reliability Corp.,

"A zebra does not change its spots."—2000 Democratic presidential candidate Al Gore—this quote from the leading political advocate of the dire effects of global warming.

"The average man I encounter all over the country regards government as sort of a great milk cow, with its head in the clouds eating air, and growing a full teat for everybody on earth."—Clarence Manion

"Al, the people we are going to visit are suffering. The president doesn't want to hear about your global warming crap."—Assistant to President Clinton speaking to Vice President Gore, who had implied the weather related event, whose victims the president was going to visit, may have been caused by global warming. As related in Roy W. Spencer's book *Climate Confusion*.

"The broad mass of a nation will more easily fall victim to a big lie than to a small one."—Adolf Hitler

"America will never be destroyed from the outside. If we falter and lose our freedoms, it will be because we destroyed ourselves."—Abraham Lincoln

"Welcome to the capitalist system. Each one of you is responsible for the amount of money you have in your pocket. The Government is not responsible for whether you eat, or whether you're poor or rich. The Government doesn't guarantee you a job or a house. You've come to a rich and powerful country, but it's up to you whether or not you continue living like you did in Cuba."—Alex Alvarez, warning Cuban immigrants what they would encounter in America.

"So you can imagine how I feel when I see the U. S. making the same mistakes that Britain has made: expanding its government, regulating private commerce, centralizing its jurisdiction, breaking the link between taxation and representation, abandoning its sovereignty. You deserve better, cousins. And we expect better."—Daniel Hannan, one of Great Britain's representatives in the European Parliament.

"Arlington (Virginia) officials boast the (federal green energy) project will save $14,000 in annual electricity costs, but the solar panels have a life span of no more than 10 to 15 years. So the feds spent $300,000 to shave at most $150,000 off the net present value of Arlington's electric bills."—Stephen Moore

"At no time shall the number of employees in the Department of Agriculture exceed the number of farmers in the United States."—Proposed amendment to the law that established the Department of Agriculture, which was laughed down by Congressional representatives. It's no laughing matter today. In 1900, at the time this amendment was proposed, there were six million farms and the Agriculture Department had 9000 employees. Today, there are less than two million farms and more than 100,000 Agriculture Department employees—and the trend continues.

"(North Dakota's economy) sticks out like a diamond in a bowl of cherry pits."—Ron Wirtz, editor of *fedgazette*, commenting on the precarious financial position of most states in 2011, principally due to underfunded public pension plans, high taxes, high unemployment and poor economies. North Dakota's unemployment rate was only 3.8%, as a result of a surging economy due to oil exploration and production, moderate taxes and its right-to-work status. 650 oil wells were drilled in 2010 and another 5,500 wells are planned over the next two decades. It was one of the few northern tier states to show an increase in population after several decades of decreases.

"Natural slaves."—Aristotle, describing those citizens who would like the government to take care of them cradle to grave.

"If you are not prepared to use force to defend civilization, then be prepared to accept barbarism."—Thomas Sowell

"Save the planet, kill yourself."—Bumper sticker, mocking the true feeling of environmentalists and climatologists towards the human race.

"The resulting famines could be catastrophic."—1975 Newsweek article forecasting a looming ice age, based on the Earth's past geologic history, since the past 11,600 years of the planet's warm interglacial period is coming to an end

"When you're picking up flack, you're probably over the target."—Unknown. Used in the context that if you're picking up a lot of criticism for something you said you probably have spoken a truth your critics don't want to hear.

"It isn't pollution that's harming the environment. It's the impurities in our air and water that are doing it."—Vice President Al Gore

"Too many religious leaders have no understanding how economies work, and thus they focus on redistributing wealth without regard for how wealth is created."—Michael Novak

"Liberty without a moral guide leads to anarchy. . . The Left implicitly admits that only a powerful state can ensure a decent society without God."—Dennis Prager from his book *Still The Best Hope: Why The World Needs American Values To Triumph.*

"The Commanding General is well aware that the forecasts are no good. However he needs them for planning purposes."—Response from a general's aide to a report by Kenneth Arrow, an Army statistician and his colleagues that the long range weather forecasts that their unit had been reporting were found to be correct only 50% of the time—a number equal to pure chance. From a report titled "Wrong Again" by Andrew Ferguson in *The Weekly Standard.*

"We have so much aluminum; it's running out of our ears."—Roosevelt administration official near the close of World War II. The Roosevelt administration chose to put the production of important wartime commodities in the hands of the country's industrialists rather than attempt to administer the economy by government decree—a decision which produced an abundance of the wartime goods needed to make America "the arsenal of democracy". Compare this experience with today's attempt by the federal government to dictate the country's healthcare system.

"No drilling or mining—not anywhere, not any time."—Unspoken goal of environmentalists.

"In general, the art of government consists of taking as much money as possible from one party of the citizens to give to the other."—Voltaire (1843) Some things never change.

"Life was nasty, brutish and short."—Unknown. This phrase has been widely used to describe the lives of the vast majority of the common people before the rise of capitalism and the industrial revolution, deemed to have started around 1820. The lives of common people improved exponentially for those countries who embraced capitalism, including those in Europe, the United States, Canada, Australia, Singapore, Japan, South Korea, and finally, the most obvious recent example—China.

"Everything that government gives to the people it must first take from the people."—Cleon Skousen

"The seductive promise of security from cradle to the grave is the real enemy of civilized society."—Alexis de Tocqueville

"If bad ideas bring horror, their antidote lies in conservative, modest, tried-and-tested ideas that respect tradition and human nature, not in revolutionary lurches and grandiose experiments, but in incremental improvements in customary practices. At a moment when many Democrats are ignoring the lessons of Venezuela and swooning over socialism, it's back to the barricades in the war of ideas."—Daniel Pipes In 1950 Venezuela enjoyed the fourth highest per capita income in the world due to its possession of the largest oil reserves on earth. Today, due to the imposition of a despotic socialist agenda its people are starving, disease and criminality are rampant, and there is mass migration out of the country. History has shown socialism to be a proven failure wherever it has been implemented, but there are still millions who are drawn to its inherent "free lunch" philosophy.

"Democratic capitalism is neither the Kingdom of God nor without sin. Yet all other known systems of political economy are worse. Such hope as we have for alleviating poverty and for removing oppressive tyranny—perhaps our last, best hope—lies in this much-despised system."—Michael Novak, who made an intellectual journey from socialism to capitalism.

"Socialism is the residue of the Judeo-Christian faith without religion. It is a belief in the goodness of the human race and paradise on earth. Capitalism is a system built on belief in human selfishness; given checks and balances, it is nearly always a smashing, scandalous success. —Theologian Michael Novak from his commentary "A Closet Capitalist Confesses", which he wrote after many years of supporting liberal and socialist causes.

"We must deflate the pretensions of self-appointed elites. These elites will hate us no matter what we do, and it is legitimate for us to help dump them into the dustbin of history."—Eric Hoffer from his book *The True Believer*

"A few students discovered that pompous teachers who catechized them about academic free speech could, with a little shove be made into dancing bears."—Adam Bloom, from his book *The Closing of the American Mind,* describing the radical students of the 60s and 70s, who took over the campuses at that time and objected vociferously against free speech, which expressed ideas in opposition to their beliefs. This intolerance arose again, with a vengeance in 21st century, regarding man-made climate change.

The climate change people have no proof of their claims. They have computer models that do not prove anything."—David Bellamy, Conservationist

"With a good conscience our only sure reward, with history the final judge of our deeds, let us go forth to lead the land we love, asking His blessing and His help, but knowing that here on earth, God's work must surely be our own."—President   John F. Kennedy, closing words of his inaugural address.

REFERENCE SOURCES

BOOKS:

Arnold, Ron and Driessen, Paul, "Cracking Big Green: To Save the Earth From the Save-the-Earth-Money-Machine", CFACT, 2014

Ball, Tim, "The Deliberate Corruption of Climate Science", Stairway Press, 2014

Berman, Bob, "The Sun's Heartbeat", Black Bay Books, 2012

Bodanis, David, "E=mc^2: A Biography of the World's Most Famous Equation", Berkley

Crockford, Susan J., Polar Bears Facts and Myths: A Science Survey for All Ages", CreateSpace Independent Publishing Platform, 2016

Driessen, Paul, "Eco-Imperialism: Green Power, Black Death", Merrill Press, 2010

Fensin, Alan, "Liars and Demons: Climate Change Truth That Anyone Can Understand", Burlington National Inc., 2015

Gray, Vincent, "The Greenhouse Delusion: A Critique of Climate Change 2001", Multi-Science Publishing Company, 2014

Idso, Craig, "Why Scientists Disagree About Global Warming: The NIPCCC Report on Scientific Consensus", The Heartland Institute, 2016

Lomborg, Bjorn, "The Skeptical Environmentalist: Measuring the Real State of the World", Cambridge University Press, 2007

Lomborg, Bjorn, "Cool It: The Skeptical Environmentalist's Guide to Global Warming", Alfred A. Knopf, Inc, 2007

Morano, Marc, The Politically Incorrect Guide to Climate Change", Regnery Publishing, 2018

Milloy, Steven J., "Scare Pollution: Why and How to Fix the EPA", 2016

Montford, A. W., The Hockey Stick Illusion", Anglosphere Books, 2015

Pielke, Jr., Roger, "The Rightful Place of Science Disasters and Climate Change", Consortium for Science Policy and Outcomes", 2014

Plimer, Ian, "Heaven and Earth: Global Warming, the Missing Science, Tyler Trade Publishing, 2009

Soloman, Lawrence, "The Deniers", Richard Vigilante Books, 2015

Spencer, Roy, "Climate Confusion: How Global Warming Hysteria Leads to Bad Science, Pandering Politicians and Misguided Policies That Hurt the Poor" Encounter Books, 2010

Spencer, Roy, "An Inconvenient Deception: How Al Gore Distorts Climate Science and Energy Policy", Amazon Digital Services, 2017

Steyn, Mark, "A Disgrace to the Profession", Stockade Books, 2015

WEBSITES:

https://forums.tesla.com/forums/100-reasons-why-climate-change-is-natural-and-not-manmade

http://zebu.uoregon.edu/~soper/Sun/fusionsteps.html (A step by step narrative explaining the sun's fusion process)

www. Wikipedia (Source for many numerical amounts used in the text

OTHER MEDIA:

Bedard, Paul, "CO_2 The 'Miracle Molecule" Key to Feeding the World", Commentary in the *Washington Examiner*, February 26, 2019

Jenkins, Jr., Holman, "Is There a Green Rational Deal?", Commentary in *The Wall Street Journal*, March 16, 2019

Milloy, Steve, "The Case For A Green 'No Deal'", Commentary in *The Wall Street Journal*, April 19, 2019

Muravchik, Joshua, "Socialism Fails Every TIme", Commentary in *The Wall Street Journal*, April 10, 2019

Portteus, Kevin, "The False Alert of Global Warming", Commentary in the *American Spectator*, July 19, 2013